ENGINEERING, SOCIAL JUSTICE, AND SUSTAINABLE COMMUNITY DEVELOPMENT

Summary of a Workshop

Rachelle Hollander, Editor
Nathan Kahl, Co-editor

Advisory Group for the Center for Engineering, Ethics, and Society

NATIONAL ACADEMY OF ENGINEERING
OF THE NATIONAL ACADEMIES

THE NATIONAL ACADEMIES PRESS
Washington, D.C.
www.nap.edu

THE NATIONAL ACADEMIES PRESS 500 Fifth Street, N.W. Washington, DC 20001

On October 2-3, 2008, the National Academy of Engineering (NAE) held a workshop on "Engineering, Social Justice, and Sustainable Community Development." This summary, prepared by NAE staff, provides summaries of the workshop presentations and discussions.

This workshop was supported in part by Contract/Grant No. 0750007 between the National Academy of Sciences and the National Science Foundation. Any opinions, findings, conclusions, or recommendations expressed in this publication are those of the author(s) and do not necessarily reflect the views of the organizations or agencies that provided support for the project.

International Standard Book Number 13: 978-0-309-15258-7
International Standard Book Number 10: 0-309-15258-5

Additional copies of this report are available from the National Academies Press, 500 Fifth Street, N.W., Lockbox 285, Washington, DC 20055; (888) 624-8373 or (202) 334-3313 (in the Washington metropolitan area); online at http://www.nap.edu.

Printed in the United States of America

THE NATIONAL ACADEMIES

Advisers to the Nation on Science, Engineering, and Medicine

The **National Academy of Sciences** is a private, nonprofit, self-perpetuating society of distinguished scholars engaged in scientific and engineering research, dedicated to the furtherance of science and technology and to their use for the general welfare. Upon the authority of the charter granted to it by the Congress in 1863, the Academy has a mandate that requires it to advise the federal government on scientific and technical matters. Dr. Ralph J. Cicerone is president of the National Academy of Sciences.

The **National Academy of Engineering** was established in 1964, under the charter of the National Academy of Sciences, as a parallel organization of outstanding engineers. It is autonomous in its administration and in the selection of its members, sharing with the National Academy of Sciences the responsibility for advising the federal government. The National Academy of Engineering also sponsors engineering programs aimed at meeting national needs, encourages education and research, and recognizes the superior achievements of engineers. Dr. Charles M. Vest is president of the National Academy of Engineering.

The **Institute of Medicine** was established in 1970 by the National Academy of Sciences to secure the services of eminent members of appropriate professions in the examination of policy matters pertaining to the health of the public. The Institute acts under the responsibility given to the National Academy of Sciences by its congressional charter to be an adviser to the federal government and, upon its own initiative, to identify issues of medical care, research, and education. Dr. Harvey V. Fineberg is president of the Institute of Medicine.

The **National Research Council** was organized by the National Academy of Sciences in 1916 to associate the broad community of science and technology with the Academy's purposes of furthering knowledge and advising the federal government. Functioning in accordance with general policies determined by the Academy, the Council has become the principal operating agency of both the National Academy of Sciences and the National Academy of Engineering in providing services to the government, the public, and the scientific and engineering communities. The Council is administered jointly by both Academies and the Institute of Medicine. Dr. Ralph J. Cicerone and Dr. Charles M. Vest are chair and vice chair, respectively, of the National Research Council.

www.national-academies.org

CENTER FOR ENGINEERING, ETHICS, AND SOCIETY ADVISORY GROUP AND WORKSHOP PLANNING COMMITTEE

JOHN F. AHEARNE, NAE, *chair*, National Academy of Engineering Center for Engineering, Ethics, and Society Advisory Group; director, Ethics Program, Sigma Xi, The Scientific Research Society

JUAN LUCENA, associate professor, Liberal Arts and International Studies, Colorado School of Mines

DEBRA STEWART, president, Council of Graduate Schools

CAROLINE WHITBECK, Elmer G. Beamer–Hubert H. Schneider professor in ethics, Case Western Reserve University

WM. A. WULF, president emeritus, NAE; university professor and AT&T professor of engineering and applied sciences, University of Virginia

Principal Support Staff

CAROL R. ARENBERG, senior editor, National Academy of Engineering

JESSICA BUONO, program associate, National Academy of Engineering Program Office (until November 2008)

CECILE GONZALEZ, program associate, National Academy of Engineering Program Office (until mid-August 2008)

RACHELLE D. HOLLANDER, director, National Academy of Engineering Center for Engineering, Ethics, and Society

NATHAN KAHL, program associate, National Academy of Engineering Program Office

JACQUELINE MARTIN, senior program assistant, National Academy of Engineering Awards Program

VALERIE HENDERSON SUMMET, Mirzayan Fellow, National Academy of Engineering Winter, 2010

Acknowledgments

This summary has been reviewed, in draft form, by individuals chosen for their diverse perspectives and technical expertise, in accordance with procedures approved by the National Academies. The purpose of this independent review process is to provide candid and critical comments to assist the committee and the National Academy of Engineering (NAE) in making its published reports as sound as possible and to ensure that the report meets institutional standards for objectivity, evidence, and responsiveness to the study charge. The reviewers' comments and the draft manuscript remain confidential to protect the integrity of the deliberative process. We wish to thank the following individuals for their reviews of this report:

John Ahearne, director, Ethics Programs, The Sigma Xi Center
Paul Citron, vice president (retired), Medtronic, Inc.
David Daniel, president, University of Texas at Dallas
Charles Ed Harris, Harry E. Bovay professor of engineering ethics, Texas A&M University
Deborah Johnson, Olsson professor of applied ethics, University of Virginia
William Oakes, director of the EPICS Program, Purdue University

Although the reviewers listed above provided many constructive comments and suggestions, they were neither asked to endorse the views expressed in the report nor did they see the final draft of the report before its public release. The review was overseen by Dr. George Bugliarello, President Emeritus and University Professor, Polytechnic Institute of New York University, who was appointed by NAE to ensure that an independent examination of this report was carried out in accordance with institutional procedures and that all review comments were carefully considered. Responsibility for the final content of this report rests entirely with the editors and the NAE.

In addition to the reviewers, the Center for Engineering, Ethics, and Society (CEES) advisory group wishes to thank the project staff. Program associates Cecile Gonzalez (until mid-August 2008), Nathan Kahl, and Jacqueline Martin managed the project's logistical and administrative needs, making sure the workshop ran efficiently and smoothly, and Jessica Buono (until November 2008) provided research and administrative services. NAE senior editor Carol R. Arenberg edited the summary drafted by CEES director Rachelle Hollander. Nathan Kahl drafted the initial reports of the small group discussions and helped edit the report and supervise the editorial and review response process. Mirzayan Fellow Valerie Henderson Summet reviewed the final draft of the summary, selected the quotes to include in the text, and worked with National Academies Press on the production of the volume. CEES director Rachelle Hollander managed the project from start to finish.

Contents

1

Introduction and Opening Remarks

The workshop on "Engineering, Social Justice, and Sustainable Community Development" was held on October 2–3, 2008, in the historic National Academy of Sciences Building in Washington, D.C. As an initiative of the Center for Engineering, Ethics, and Society at the National Academy of Engineering (NAE CEES), the workshop was cosponsored by the Association for Practical and Professional Ethics, with support from the National Science Foundation and NAE member Harry E. Bovay Jr., the underwriter of activities of the CEES.

"Engineering, Social Justice, and Sustainable Community Development" is the first in a series of biennial workshops planned by CEES on the theme of engineering ethics and engineering leadership. This workshop was inspired by members of the CEES Advisory Group (CEES-AG), who raised questions about conflicting positive goals for engineering projects in impoverished areas and areas in crisis. These conflicts arise domestically as well as in international arenas. CEES-AG noted that engineers and ethicists had not examined or discussed the difficulties such conflicts could pose for successful project completion. These goals of project sponsors and participants, which are often implicit, include protecting human welfare, ensuring social justice, and striving for environmental sustainability alongside the more often explicit goal of economic development or progress.

At the first meeting of CEES-AG in November 2007, the group (and CEES director) agreed on the focus for the workshop and developed an agenda and a list of potential participants. The group also established the following ambitious goals for the workshop:

- Improve research in engineering ethics.
- Improve engineering practice in situations of crisis and conflict.
- Improve engineering education in ethics and social issues.
- Involve professional societies in these efforts.

In pursuit of these goals, the workshop agenda included panel sessions, a roundtable, small group discussions, and plenary reports. There was also a planning session on the evening prior to the event and a wrap-up session immediately afterward that included members of the advisory group, session moderators, and leaders and rapporteurs of the small discussion groups. Follow-up plans include presentations at professional meetings and publications, including the present workshop summary. The workshop agenda can be found in Appendix A. The workshop planners, program participants, and attendees are listed in Appendix B.

The main body of this summary highlights presentations and discussions by the panels and roundtable. It also includes highlights from the reports of the small group discussions. Papers and PowerPoint slides submitted by program participants are available on the CEES website (*www.nae.edu/ethicscenter/*).

1.1 OPENING SESSION

NAE member John Ahearne, chair of CEES-AG and director of the Ethics Program, Sigma Xi, moderated the opening session. NAE President Charles M. Vest welcomed the participants, and former NAE President William A. Wulf, who is also a member of CEES-AG, introduced CEES and the workshop program.

In his remarks, Dr. Vest emphasized the intricate connections between science and engineering and larger social, political, and economic systems. As the complexity of these interactions increases, he said, they can geometrically increase the difficulties of determining the right thing to do and determining the pathways for achieving sought-after solutions. Dr. Vest pointed out the need for engineers to adopt a global vision, citing examples, such as the moratorium on recombinant DNA research, the Montreal protocols on chlorofluorocarbons in the atmosphere, and the establishment of the World Wide Web consortium, as indicative of science and engineering initiatives that have reflected sensitivity to societal needs and goals.

Dr. Wulf then set the stage for the workshop. The great challenge for engineers in the 21st century, he said, will be addressing the ethical issues and responsibilities associated with increasingly complex systems and interactions. He recalled that in his presidential address to NAE in 2000, he had thought that "the appropriate thing to do would be to talk about the accomplishments of engineering for the last 100 years and the challenges in engineering for the coming 100 years." Engineering achievements in the 20th century had brought people enormous benefits and some burdens, most of which had not been predicted and, perhaps, were not predictable. He said the complexity Dr. Vest had highlighted in his remarks demonstrates the ethical challenges ahead, and responding to those challenges in the context of increasingly powerful, potentially irreversible[1] engineering innovations is the major issue facing the engineering profession.

Dr. Wulf described how he had established an advisory committee during his presidency to consider what NAE could do to meet this ethical challenge. That committee had provided a report that was the basis for establishing CEES, with core funding provided by NAE member Harry E. Bovay, Jr. Thus, he said, we have arrived at this first workshop in a hoped-for biennial series on engineering ethics and engineering leadership. The diversity of the audience and the workshop program, he said, indicates a successful beginning.

Dr. Ahearne and CEES Director Rachelle Hollander then reviewed the agenda and purposes of the workshop and introduced the first panel.

[1] Dr. Wulf has used examples from computing and ecology to make this point. Complexity in computer software means that changes in properties cannot always be predicted. Engineering natural systems such as the Everglades results in irreversible change; see his keynote address in National Academy of Engineering, 2004, Emerging Technologies and Ethical Issues in Engineering, The National Academies Press, Washington, DC.

2

Session I: Engineering and Special Vulnerabilities

The focus of Session I was summed up in the following statement:

> Engineers and engineering organizations operate in circumstances of crisis, ranging from conflict to disaster. They operate where human rights problems are highly visible and where issues of sustainable community development arise. This session reports perceptions about the technical and social constraints and opportunities they face and whether and how aims for humanitarian action, social justice, and sustainable community development can be met.

This session, which was moderated by National Academy of Engineering (NAE) member Henry J. Hatch, former Commander, U.S. Army Corps of Engineers, included three presentations and two responses. The first talk was given by Abul Barkat, Department of Economics, University of Dhaka, Bangladesh, and Abul Hussam, Department of Chemistry, George Mason University, and winner of the NAE Grainger Challenge Prize for Sustainability. The subject was the human rights challenge of providing arsenic-free drinking water in Bangladesh. In the second talk, Christopher Seremet, Technical Advisor, Water Supply and Sanitation, Catholic Relief Services, described challenges facing engineers operating in crisis conditions, particularly if they are not familiar with the political and social context of the project. Finally, Anu Ramaswami, Department of Civil Engineering, University of Colorado, Denver, discussed challenges to addressing chronic problems in ways that ensure sustainable community development. Discussants were NAE Foreign Secretary George Bugliarello, president emeritus and university professor, Polytechnic Institute of New York University; and Deborah Goodings, professor, Engineering and Public Policy, University of Maryland, College Park.

2.1 REMOVING ARSENIC FROM DRINKING WATER IN BANGLADESH

In the opening presentation, doctors Abul Barkat and Abul Hussam described the stark human rights and public health situation in Bangladesh. Official estimates indicate that 50 percent of the population of some 150 million is at some risk of arsenic poisoning from groundwater (used for drinking) from tube wells. Thirty-five percent have no access to arsenic-free drinking water and thus cannot avoid this risk. In this dire situation, poorer households are at significantly higher risk, and arsenicosis (a disease caused by arsenic poisoning) leads to even deeper poverty and more social stigma.

Dr. Hussam explained how the SONO filter, which he developed, addresses these issues. There were significant difficulties—financial, technical, and social—in introducing a system such as the filters. Providing six million filters to the poor would cost U.S. $300 million, or 10 percent of the annual public improvements development budget of Bangladesh. In addition, the production, distribution, and necessary follow-up on the use of filters pose unprecedented management challenges for the country. Social issues include: lack of awareness among social gatekeepers, such as influential media outlets and employers; widely held myths and misconceptions among the populace; and a lack of commitment by the government.

Nevertheless, both speakers agreed that there are opportunities for improving management capacity, developing local solutions, and providing public education about safe drinking water. Field tests have demonstrated the viability of an integrated arsenic-mitigation program that includes external development partners; the government of Bangladesh; electronic media; civil society; nongovernmental organizations (NGOs); and schools that recognize and accept the imperative of producing, deploying, and using SONO filters to increase access to safe water and thus improve public health and nutrition.

2.2 ENGINEERING PROJECTS IN INTERNATIONAL EMERGENCIES

Christopher Seremet, technical advisor, Water Supply and Sanitation, Catholic Relief Services, focused his remarks on the constraints facing engineers working in international emergencies, which require an immediate response, to either mitigate or prevent catastrophe. Some emergencies, such as earthquakes and industrial accidents, have a very quick onset. Others, such as droughts, develop over a period of time. In either case, however, an intolerable situation demands an emergency response.

> "[I]nternational emergencies often involve life or death situations and this can be. . . overwhelming for a young engineer. . . . Information comes in slowly, critical decisions have to be made immediately. . . . You're out of your zone from a traditional engineering standpoint. . . ."
>
> Christopher Seremet,
> Catholic Relief Services

In emergency situations, constraints over human rights and social justice affect technological decisions, he said, and engineers must be prepared to operate beyond their normal comfort zone. They will be forced to deal with situations for which traditional educational programs have not prepared them, he said, and they must put aside deliberate, thorough processes of data collection, analysis, and planning and identify problems and formulate solutions quickly, often during the initial visit to the site. These constraints raise substantial concerns, he said, particularly for inexperienced engineers.

Decisions made under these circumstances, which can have life-or-death implications, may have to be based on inadequate information, rapidly changing conditions, and a very slow rebuilding process. Such situations call for empathy and other "people skills" as much as, or perhaps even more than, engineering skills. It might even be difficult to identify an effective engineering solution. War or civil conflicts, which arouse anger, suspicion, and distrust, pose additional challenges. There may also be issues of corruption and misuse (stolen equipment or money; bribery), and, in some cases, emergency response personnel have even been subject to violence.

On a more personal level, engineers must consider the potentially high costs to themselves of undertaking an emergency relief assignment—in terms of prior commitments, long-term career goals, and other factors.

2.3 ADDRESSING THE SUSTAINABILITY CHALLENGE

In her talk, Anu Ramaswami, Department of Civil Engineering, University of Colorado, Denver, laid out some broad-based principles for learning to address the challenge of sustainability in responding to chronic, rather than emergency, problems. One principle for addressing the issues raised in this workshop is: "Develop and apply engineering solutions, while being cognizant of local geography, aspirations, and cultures." Dr. Ramaswami focused on "how to do that, how to do it as an outsider, and how to teach engineers and engineering students to be able to do it."

She recommended that three questions be asked when this kind of project is undertaken: Who owns the project or sets the agenda? Whose knowledge counts in the design? Who benefits? As she explained, project selection may be under the control of particular stakeholders, even though others may have claims that are just as strong, or even stronger. The answer to the question of who owns the project or sets the agenda will, of course, affect the answers to the next two questions.

In answering the second and third questions, she reminded listeners, unless local knowledge is taken into account, the outcome is not likely to be sustainable. Even if a project has an appropriate goal, one may look back and find that the intended beneficiaries had not been served.

She concluded by noting that what is needed in engineering education is an integration of expertise and techniques from the social sciences so that engineering projects are better defined. Specifically, engineering can take from the social sciences the understanding of community-based, participatory research. Planners must ask the important questions, "Who is the project for? Whose knowledge counts? Who takes part in . . . problem definition, data collection, interpretation, and analysis? Who takes action? Who owns the project? And how are the results assimilated?"

2.4 PANEL DISCUSSION

The panel then identified issues for discussion based on the presentations. George Bugliarello, Polytechnic Institute of New York University, pointed out that all three presentations had shown that the problems facing engineers are as much social as technical. The situation in Bangladesh demonstrated first how poverty creates disease and social stigma, which, in turn, increase poverty, and second that scientific and technological solutions to problems require an organizational response to succeed. The second talk, on engineering in a crisis, he said, posed challenges to engineering education, such as how we can prepare students to act in crisis situations where quick decisions are necessary. The talk on addressing chronic, long-standing challenges showed the importance of honoring general principles to ensure that relevant social groups are included in setting the agenda.

Deborah Goodings, University of Maryland, focused her remarks on the desire of engineers and engineering students to be doing, and not just talking. Given the difficulties the speakers had identified, she asked for their recommendations for promoting the effective involvement of engineers and engineering students in local community and international development projects and for broadening engineering education to address the complex problems these projects and circumstances raise.

2.5 GENERAL DISCUSSION

In the general discussion that followed, attendees identified a number of important issues. Richard Anderson of SOMAT Engineering, Inc., pointed out that planners should take into consideration what the "added value" is, and for whom, when engineering students from the United States or other developed countries undertake projects in developing countries. Legitimizing these efforts requires measuring positive results—for student participants and for the community. Rebekah Green, Western Washington University, said that measuring positive results must also include what happens in the future. Are engineering solutions sustainable in the long term?

NAE member Alice Agogino, University of California at Berkeley, noted that the deep tube wells in Bangladesh, which had been dug to provide abundant drinking water to large numbers of people, had unexpectedly caused the arsenic problem, demonstrating the need for whole systems analysis in deciding whether to implement an engineering solution to a problem. Systems analysis, she said, can help engineers avoid, or at least minimize, negative consequences from their interventions. In response, Dr. Barkat pointed out the difficulty of isolating a single source for the problem of arsenic in the water. He emphasized the need for better systems-oriented education to enable engineers to at least recognize, if not address, the complexities of such problems.

David Crocker, Institute for Philosophy and Public Policy, University of Maryland, College Park, noted that engineers must wrestle with underlying ethical questions, such as who *should* own a project, whose knowledge *should* count, and who *should* benefit. For example, if a solution will benefit males but not females, how should an outsider proceed? What are the responsibilities of each party? In response, Anu Ramaswami explained that NGOs are developing codes of ethics that might help answer questions about responsibility. She suggested that end-point assessments include community development as well as technical criteria.

Moderator Hatch ended the session by repeating that engineering in context requires attention to a great many factors: "technical, environmental, social, political, cultural, ethical, and you could probably add a whole bunch of others." Unless all of these factors are taken into account, engineering efforts will fail, he said. He then asked that academics help practicing engineers to succeed. He pointed out how few practicing engineers had attended this meeting and recommended that greater efforts be made to engage them in these kinds of discussions.

3

Session II: Engineering, Ethics, and Society

The purpose of this session was to explore interfaces between engineering, ethics, and practice in circumstances of conflict, poverty, or emergency:

> Do humanities and social sciences disciplines bear on problems for engineers and engineering professions in such circumstances? Has the field of engineering ethics drawn adequately from this scholarship or the real exigencies of engineering practice? Presenters were asked to "examine technical, political, historical, environmental, economic, and cultural constraints that shape outcomes."

David Crocker, Institute for Philosophy and Public Policy, University of Maryland, College Park, moderated the session. The presenters were Ronald Kline, professor, Science and Technology Studies, Cornell University; Carl Mitcham, professor, Department of Philosophy, Colorado School of Mines; and Wesley Shrum, professor, Department of Sociology, Louisiana State University. Discussants were Priscilla Nelson, provost and vice president for academic affairs, New Jersey Institute of Technology, and Donna Riley, professor, Picker Engineering Program, Smith College.

3.1 INTEGRATING MACRO-ETHICS AND MICRO-ETHICS

Ron Kline, Cornell University, kicked off the session with a talk on the use of case studies to teach engineering ethics and suggestions for how general social concerns about technology (often called "macro-ethics") can be integrated into the agent-centered approach ("micro-ethics"). He noted that the purpose of using case studies is to place students in decision-making roles that resemble those faced on the job.

Case studies can have many different characteristics, he said. They may be historical or hypothetical, provide positive or negative role models, provide a focus on everyday or rare events, or on individual or organizational actions. They may require prospective or retrospective analysis, describe problems of conflicting values or "where to draw the line," be very sketchy or very intensively described, or focus on an issue of professional conduct or of technology and society. But very often they reduce complexity to an individual choice rather than requiring an analysis of "interactions between individual actions and organizational responsibilities," including public policy. He noted that in the workplace engineers do the research on a case themselves and write their own case studies, of a sort, in the process of making decisions. Thus he suggested that engineering students might learn more by creating their own case studies as they may have to do eventually in their jobs.

He acknowledged that many case studies could be used to highlight issues of collective responsibility in major social systems, which necessarily include scientific and technological components, such as the classic case of the investigation into the crash of a DC-10 in Paris in 1974. Cases illustrating the involvement of international corporations, unions, airports, and government agencies would lead to more exploring of such issues, but most cases have been boiled down to individual decisions. An integrated approach that includes macro-ethics would also have examined actions of and interactions between engineering professional societies, legislative and executive government agencies, and corporations and unions, including designated engineering representatives nominated by manufacturers and trained by the Federal Aviation Administration.

Dr. Kline suggested that investigations of engineering design cases involving environmental issues naturally include widespread social concerns. He is currently looking into using the Lake Source Cooling Project at Cornell University (an environmentally friendly redesign of Cornell's chilled water system, eliminating refrigeration and its associated energy use) as a teaching example.

3.2 HUMANITARIAN ENGINEERING

Carl Mitcham, Department of Philosophy, Colorado School of Mines, began by noting that involvement of the humanities and social sciences in engineering ethics and practice has been attenuated, partly, he believes, because those disciplines do not yet have a significant body of scholarship bridging theories and outcomes from which engineering can draw, though they do have things to contribute. Historically, he said, engineering as a profession has been influenced more by external rather than internal values. In 1828, English railway engineer and writer Thomas Tredgold said, "Engineering is the art of directing the great sources of power in nature for the use and convenience of man." Today, the emphasis has changed from "use and convenience" to "public safety, health, and welfare," but these are the definitions of non-engineers, he said.

Humanitarian engineering involves critical reflection on public safety, health, and welfare, and so engineering and humanitarianism have parallel histories and parallel interests. Their joint history can be traced back from nursing and the Red Cross in the mid-1800s, to engineering relief work during World War I, the many international institutions during World War II that have since incorporated engineering activities, and individual engineers with distinguished records of humanitarian work. A reflection of this progression is the establishment of Engineers Without Borders in the 1990s. In academia, Dr. Mitcham noted, courses in humanitarian engineering and engineering and sustainable community development were taught on his own campus and elsewhere.

> "[T]he current code of professional conduct of most engineering societies emphasizes . . . public safety, health, and welfare. But what are public safety, health, and welfare? Engineers don't study that."
>
> Carl Mitcham,
> Colorado School of Mines

Dr. Mitcham ended his presentation with challenges to engineers:

- Engineering practice makes assumptions about the beneficence of engineering that deserve critical examination—to which the humanitarian movement can contribute.
- Engineers can make crucial contributions to humanitarian relief efforts—especially if they are self-critical about what it means to be an engineer.
- The humanitarian engineering possibility may be able to help address the pipeline problems.
- Humanitarian engineering is simply one of many efforts to broaden and transform engineering.

He ended with a quote from *The Engineer of 2020: Visions of Engineering in the New Century* (National Academy of Engineering, 2004)[1]: "We aspire to a future where engineers are prepared to adapt to changes in global forces and trends and to ethically assist the world in creating a balance in the standards of living for developing and developed countries alike."

3.3 REAL-WORLD ETHICAL DEBATES

Wesley Shrum, Louisiana State University, began by acknowledging that scholars in science and technology studies (STS) have been reluctant to engage in ethical debates except in hypothetical contexts. He paid tribute to the commitment of engineers to examining and analyzing the infrastructure failures during Hurricane Katrina in New Orleans, even as the rebuilding began. Five major teams, funded by a variety of sources, were actively involved, some collecting data and some reviewing records of the events.

Dr. Shrum himself was a member of a team, but he resigned to pursue his work in sociology from a more neutral position. His research—analyses of interviews with the engineers involved on the panels—is expected to take 10 years. He pointed out that the teams have become mired in controversy and accusations, for instance, of cover-ups or malfeasance; that teams have merged and separated; and that suspicions and tensions, charges and countercharges have arisen. In addition, each group has struggled to agree on what to say, when to say it, and to whom and by whom it should be said. These are the kinds of questions, he said, that STS should be addressing and that STS scholars should assist practitioners to address.

On another level, accusations of malfeasance against one team resulted in the establishment of a committee by the American Society of Civil Engineers to address issues related to the process, funding, communication to the public, and conflicts of interest related to engineering reviews. As secretary of the Society for Social Studies of Science (4S), Dr. Shrum believes that these issues too should be addressed by STS scholars. For example, 4S should take positions based on the results of research by its scholars on who should be members of technical committees and on when and how and what they should communicate to

[1]National Academy of Engineering, 2004, The Engineer of 2020: Visions of Engineering in the New Century, The National Academies Press, Washington, DC.

the public; 4S should also address conflicts of interest on engineering review committees.

3.4 PANEL DISCUSSIONS

Priscilla Nelson, New Jersey Institute of Technology, drew on her own work in rapid-response engineering and on the development of linkages between the National Science Foundation Directorate for Engineering and the Directorate on Social, Behavioral, and Economic Sciences. She argued that integrating engineering and social sciences is critical to the future of engineering. The profession must have the ethical right to learn from disasters and other rare but anticipated events, she said. With that ability, the engineering profession could move from a data-poor state to a data-rich state. It could plan for postmortems and develop the data resources necessary to validate simulations that are now limited to hypothetical and speculative results that are open to challenge. She argued that case-based learning should be integrated throughout the engineering curriculum to ensure that young engineers are sensitive to the problems entailed in mega-scale projects. She said students should be recruited with the expectation of becoming practitioners of "a profession of integrated solutions."

Dr. Shrum interjected that risk is central to the future of engineering, which means that the role of insurance and cultural perceptions must also be integrated into engineering education.

> "As educators I think we need to challenge students of engineering ethics at the level of problem definition; we need to challenge them to think critically about engineering as well as at the same time they think critically within engineering."
>
> Donna Riley, Smith College

The other discussant, Donna Riley of Smith College, addressed essential prerequisites for engineers working toward social justice. As an engineering educator, she said that students would have to let go of their "absolute faith in engineering analysis as objective truth" and pay much more attention to context. In a challenge to the audience, she asked that everyone consider whether sociologist Robert Zussman was correct in his belief that engineers were taught not to question the motives or incentives of others, but to accept them and proceed to calculate the means of achieving them. If this is so, Zussman concluded, engineering is not a profession.

Dr. Riley argued that for engineers to be professionals, and to be considered as such, they must exercise individual autonomy and collective responsibility in working for social justice. She acknowledged, however, that exercising that responsibility would be difficult because engineers are not autonomous. Instead, she said, they are in "social captivity" insofar as they work in non-executive capacities for large corporate or government institutions. For the work of engineers to advance social justice, she said, they must be aware of the particular circumstances of each project, including the historical context, negative impacts of globalization, racism, classism, and sexism.

3.5 GENERAL DISCUSSION

Jonathan Herz, an architect in the U.S. General Services Administration, opened the discussion by restating that engineers have a responsibility for defining problems when environmental or human stakes are high and correcting the negative unintended consequences of past engineering mistakes. Aarne Vesilind, retired professor of civil engineering, Bucknell University, pointed out that "real life" requires engineers to extract significant information from obscure situations. However, he said, engineering students are not accustomed to, and sometimes even resent, being presented with problems that simulate such contexts. Providing information of varying degrees of relevance might improve engineers' ability to identify or extract important or missing information in engineering analyses.

NAE member Alice Agogino and Dr. Riley then mentioned the role of feminist scholars who have raised questions about traditional power hierarchies, even in classrooms. Engineering students are not taught to consider that who poses a question (usually the more powerful party) may shape that question in a way that predisposes the answer to be most acceptable or helpful to the questioner. Feminist approaches to pedagogy bring these questions out into the open. Even the power of the teacher in the classroom has been questioned from this perspective. Dr. Nelson then suggested that power and setting agendas for engineering research, education, and practice might also be addressed in continuing professional education.

Dr. Shrum's decision to leave the Katrina study team was questioned by a participant. He responded that his expertise is in documenting the underlying social and organizational conditions of conflicts and their role in engineering ethics, rather than in answering engineering questions. In addition, he had concluded that his effectiveness depended on his research subjects being confident of his neutrality.

CEES-AG member Caroline Whitbeck, retired professor in ethics, Case Western Reserve University, and founder of the Online Ethics Center, took issue with the argument that engineers are less self-directed than physicians. In response, Dr. Mitcham argued that, although he knew this idea was contentious, he thought everyone would agree that a good deal could be learned from the humanitarian tradition that would improve the engineering profession and engineering ethics.

4

Panel on Early-Career Engineers

After lunch, Rebekah Green, Institute for Global and Community Resilience, Western Washington University, and Daniele Lantagne, Centers for Disease Control (CDC), addressed the group as members of a panel moderated by Joseph Le Doux, Department of Biomedical Engineering, Georgia Institute of Technology. Both women are junior, or early-career, engineers who have worked on social justice and sustainable community development. In their talks, they addressed the following questions:

> What led you to your career choices? How do you see them in relationship to the goals of this meeting, to enhance engineering research and practice and improve engineering education through attention to issues of engineering, social justice, and sustainable community development?

4.1 "COMMUNITY ENGINEERING"

Dr. Green described the choices that led her to enter the field of "disaster research," which requires combining scientific, engineering, and community knowledge. After earning an undergraduate degree in civil engineering and several years working in a structural design firm, Dr. Green decided to do graduate work in Cornell University's Civil Engineering Department. Her advisor allowed her to take courses in anthropology and science and technology studies, stipulating that she had to integrate the work into her dissertation research. She was able to do this by specializing in disaster studies.

In the research for her dissertation, she examined perceptions of seismic risk and building decisions in squatter settlements that comprise about 50 percent of the built environment in and around Istanbul, Turkey. She found that perceptions of physical and social risk influenced building decisions. People in the settlements did not trust engineers and insisted on constructing buildings without engineering assistance. By working with a public education unit in Istanbul that included engineers, construction workers, sociologists, and social workers, Dr. Green developed a program to teach the public about good building to diminish seismic risk. Despite some negative reactions, the program has been integrated into trade schools and the education ministry.

> "I am perceived as something that is not an engineer. I'm working there as a disaster researcher. . . . [My] engineering background is just sort of this interesting personal history . . . but I feel like more of an engineer than ever because I'm actually working towards innovating for the future which is what engineers do."
>
> Rebekah Green,
> Western Washington University

Dr. Green said other projects she undertook in Central Asia were not as successful, because the stakeholders could not agree on minimally acceptable risk. In all of the projects, she said, "we weren't just debating technical details but professional responsibility. It was an attempt to ensure safety in an area of extremely high uncertainty. [T]he result was [not only a] compromise on technical details but also a compromise on ethical principles. . . . [However,] we all had to agree that these guidelines were better than nothing." Understanding the social context in which the construction would be done was essential to ensuring that local construction companies and workers would proceed in a safer fashion than they had previously, although the new construction might not satisfy the standards taught to U.S. engineers.

Dr. Green also described her work in post-Katrina New Orleans on damage assessment and recovery plans for a low-income neighborhood. Working with 80 students from Cornell, Columbia, and the University of Illinois, Urbana, she helped develop a plan for the lower Ninth Ward that was supported by the New Orleans City Council.

Afterward, when she began looking for an academic post, she found that her priorities did not match those of traditional engineering departments, which considered sustainability and humanitarian work as complementary, rather than essential, to their programs. She now teaches in a program that trains urban planners, disaster mitigation specialists, and emergency planners, but not engineers. Nevertheless, she believes that as a "community engineer" her research and advice, although less valued by engineers and anthropologists, may make a significant contribution to the creation of a just society.

4.2 ENVIRONMENTAL ENGINEERING

Ms. Lantagne, CDC, described environmental engineering as a critical component of public health. Her career began with her interest in protecting rivers. With a bachelor's degree from Massachusetts Institute of Technology (MIT) in environmental engineering, she was able to use her technical engineering skills to promote sustainable watershed communities in Massachusetts. Her master's degree, also from MIT, focused on household water treatment and involved courses at the John F. Kennedy School of Government (Kennedy School) at Harvard University. In her master's thesis, an examination of trihalomethane levels in a double-bucket filter system used in Haiti indicated that the system was safe. She subsequently has received her qualification for professional engineering in environmental engineering.

At CDC, she provides technical assistance to nongovernmental organizations (NGOs) working in developing countries to improve water in many ways. In the past eight years, she has worked in more than 40 countries, sometimes involving a brief trip to a site, a chemical test, a quick intervention, or consultation with follow-up from local stakeholders. For instance, a small CDC team may help with water-quality testing, or product development or selection, or government approvals. Ms. Lantagne used the example of the importance of treatment of household water. The lack of properly treated water, she said, contributes greatly to the deaths of children under the age of five in many areas of the world. We have the technologies to solve this problem, she said, but the difficulty is getting them to the people who need them.

It's important, she noted, to recognize that people with different roles describe the problem in different ways. Water must be safe to drink in a user's cup or hands. Just because it leaves a water-treatment plant free of fecal coliform, which may be satisfactory to plant engineers, is not sufficient to ensure its safety as drinking water. The same is true of delivering water through a pipe. Ensuring the safety of drinking water will require the political will and economic planning to improve the water supply, water treatment, latrine building, and hand-washing behaviors. Technological verification of water quality is necessary, but much, much more is required for a program to be successful.

Ms. Lantagne argued that effective communication is essential to

> "We asked a woman in a rural home to give us a glass of water the way she would prepare it for a child and we test that."
>
> Daniele Lantagne, CDC

changing habitual behaviors. A successful project requires integrating skills and knowledge from several academic fields to implement and monitor a sustainable system that can be scaled to meet community needs. In Kenya, for example, local companies prepare and bottle a very low-cost chlorine solution distributed by an international NGO marketed by radio, TV, and street theater. Midwives who use the solution recommend it to new mothers.

4.3 DISCUSSION

The discussion that followed was focused on identifying the components of an engineering curriculum or activities that would encourage students to develop a better understanding of ethical dimensions and subtleties. As the two women had said, Ms. Lantagne's undergraduate schooling had included and encouraged both engineering content and courses outside engineering that had contributed to her approach to solving complex, poorly defined problems. Dr. Green's schooling had not included those "outside" influences. Nevertheless, Dr. Green believes that considerations of sustainability are more encouraged today than they were when she was a student.

Environmental engineer Jonathan Essoka, Environmental Protection Agency, Philadelphia, commented that attracting students from diverse backgrounds to engineering could create a climate for more diversity in engineering education. Ms. Lantagne agreed and pointed out that the National Academy of Engineering website, *EngineerGirl!*, which emphasizes that the purpose of engineering is to enable young girls to become what they want to be—"whether that's environmental justice in the States or in developing countries or building a cool spaceship, we need to do all of it," sends a strong message about the potential social and humanitarian benefits of engineering. Dr. Agogino added that funding agencies should take the societal effects of proposed engineering research projects into consideration in their funding decisions.

5

Session III: Implications for Engineering Education

In this session, moderated by Woodie Flowers, National Academy of Engineering (NAE) Member, Department of Mechanical Engineering, Massachusetts Institute of Technology, workshop participants were asked to consider the previous discussions in the context of engineering education:

> Engineering educators and professional societies can promote attention to engineering in circumstances of social conflict and environmental challenge, and prepare students and members to address issues responsibly. What kinds of challenges do such activities bring to engineering education? How are educational and professional programs responding? What structural, programmatic, and/or curricular changes, if any, are underway? What alliances are needed?

The presenters were NAE member Linda Abriola, dean, School of Engineering, Tufts University; Caroline Baillie, Department of Materials Engineering and Engineering Education, Queens University, Canada; and Kevin Passino, Department of Electrical and Computer Engineering, Ohio State University. Discussants were Richard Anderson, SOMAT Engineering, Inc., and David Daniel, NAE member and president, University of Texas, Dallas.

In his introduction of the session, Dr. Flowers mentioned FIRST (For Inspiration and Recognition of Science and Technology), an organization he works with that sponsors competitions in robotics for young people. The underlying big idea of FIRST is "gracious professionalism," he said, and the biggest award is for teamwork and professional conduct.

5.1 ENGINEERING PROGRAMS AT TUFTS UNIVERSITY

Kick-off speaker Linda Abriola presented an overview of Tufts programs, which promote "the education of engineers as active citizens and innovative problem solvers," and highlighted "common elements, strengths, and challenges" of the educational models on which they are based. The School of Engineering, in cooperation with the Tisch College of Citizenship and Public Service and the Tufts Institute for the Environment, uses "project-based service learning models" in its undergraduate courses to encourage innovation, teamwork, cross-disciplinary collaboration, and leadership. "Curricular models range from mentored senior projects in sustainable design to a university-wide seminar in the Institute of Global Leadership."

Cross-school programs highlighting specific topics (e.g., water; or systems, science, and society) encourage interdisciplinary collaborations in graduate engineering education. These programs often include participants from the social sciences and other science disciplines. Extracurricular activities, which are strongly encouraged, include participation in national organizations, such as Engineers Without Borders, and membership in Tufts outreach groups, such as NERD Girls[1] and STOMP (Student Teacher Outreach Mentorship Program) of the Center for Engineering Educational Outreach.

Dr. Abriola listed the following factors as important to the success of these programs: institutional commitment; alignment of university culture; supportive administrative infrastructure; individual leadership; and assessment, feedback, and dissemination. These factors are also highlighted in the university's mission statements.

5.2 ENGINEERING WITH AN EYE TO SOCIAL JUSTICE

Caroline Baillie, Department of Materials Engineering and Engineering Education, Queens University, Canada, focused on the characteristics of programs that educate engineering students with an eye to social justice. In her presentation, she (1) identified the main principles of socially just engineering practice and (2) discussed the potential of putting social justice at the center of engineering education and practice. Both of these can show the way to transforming engineering education.

[1]"Nerd" is a term often used to describe "eggheads," or intellectually and technically astute youngsters who are perceived, stereotypically, to lack social skills. See *http://www.nerdgirls.org/About.html.*

The first principle of socially just engineering practice, Dr. Baillie said, is that engineering students must be educated to think critically and to question perceived views; this principle is based on Paulo Friere's definition of critical thinking as the ability to "see the world as changing, rather than as static," a place where technological needs arise in "diverse social, political, and economic contexts." The next principle is recognizing that "in an academic context, stating opinions is less important than taking positions." Taking a position requires understanding the frames of reference and thinking through the values at issue. Only then can social justice, rather than charity, become a consideration.

> "How do we get students to move . . . out of their comfort zones and actually critique the positions that they have always been told about from their parents, from their schools, and what they hear [in the media]? That's not an easy thing but it's really critical."
>
> Caroline Baillie,
> Queen's University

Dr. Baillie described several methodological approaches to helping students master these skills: "threshold concepts, action research, collaborative enquiry, transformational learning, and phenomenography."[2] She and her collaborators have developed learning modules combining various technical and social issues, such as thermodynamics and social justice (issues related to continual growth and resource exploitation). A network addressing these issues and questions can be accessed at *esjp.wikispaces.com*.

5.3 EDUCATING VOLUNTEER ENGINEERS

Kevin Passino, Ohio State University, took a position in his talk—that educating volunteer engineers is a university responsibility and that fulfilling that responsibility requires several strategies: putting more emphasis on ethics and professionalism in the curriculum; encouraging hands-on volunteerism via student organizations; and promoting

[2] "Phenomenography is the empirical study of the differing ways in which people experience, perceive, apprehend, understand, [and] conceptualize various phenomena in and aspects of the world around us." Researchers in phenomenography investigate and classify the different ways people learn. See Ference Marton. 1994. Phenomenography. Pp. 4424–4429 in *The International Encyclopedia of Education*, vol. 8, 2nd ed., edited by T. Husén and T. N. Postlethwaite. New York: Pergamon Press. Available online at *http://www.ped.gu.se/biorn/phgraph/civil/main/1res.appr.html.*

service learning through community-oriented design projects. However, developing the academic infrastructures that can encourage and support engineering volunteerism is a significant challenge.

In support of his position, Dr. Passino noted that the definition of a profession always includes public service. This doesn't mean that every engineer has to satisfy this criterion, he said, but *the profession as a whole* does. He then gave some examples of class assignments that could "teach" ethics/professionalism, such as paper design projects that must meet community design constraints or address global issues, and research papers on relevant subjects, such as assessing corporate citizenship programs, surveying engineering volunteerism projects, assessing codes of ethics, and so on.

Dr. Passino then described ECOS (Engineers for Community Service), a student-run organization at Ohio State that links students with sponsors of local and international service projects that promote professionalism.[3]

> "Public service is a crucial part of being a professional."
>
> Kevin Passino,
> Ohio State University

Finally, Dr. Passino argued for an infrastructure that goes beyond academia and involves professional organizations and government, as well as industry, where corporate citizenship programs could have a big impact.

5.4 PANEL DISCUSSION

In response to the talks in this session, Richard Anderson of SOMAT Engineering, Inc. described what a few professional societies are doing to encourage socially just and environmentally sensitive engineering, as well as to promote professionalism and ethics. ABET[4] is working with China to develop an engineering accreditation system, an ethical obligation, he argued, to help developing nations. Dr. Anderson also mentioned work by the American Association for Engineering Societies, an umbrella organization that represents the United States in the World Federation of Engineering Organizations, which has a committee on capacity building.

[3]For more about the ECOS-sponsored activities, see *ecos.osu.edu* for project descriptions.

[4]ABET, Inc., the recognized accrediting agency for college and university programs in applied science, computing, engineering, and technology, is a federation of 29 professional and technical societies in these fields. See *www.abet.org*.

Dr. Anderson then turned to issues raised in the presentation by Dr. Green, the early-career engineer. He described the contentious issues in post-Katrina New Orleans that arose when the recommendations of engineering experts conflicted with community needs and desires, such as a recommendation against rebuilding on a floodplain. He left the group with this question: Is this an example of a conflict between professional ethics and social justice? Can engineering education begin to examine and analyze professional responsibility in light of such conflict?

The second discussant, David Daniel, NAE member and president, University of Texas, Dallas, said that licensed professional engineers had "built a box around themselves" by saying who can and cannot be an engineer. On the one hand, this protects them from incompetent engineering. On the other hand, it may prevent professional engineers and engineering faculty from moving forward and from "intellectually permeating out into some of the softer boundaries on the periphery of the society."

> "I see . . . a lack of willingness of a lot of faculty . . . to expand the size of this box that we call engineering."
>
> David Daniel,
> University of Texas, Dallas

Dr. Daniel has argued that supporting courses for master's degrees should not be taught in the college of engineering, a point of view that "appalls" many of his colleagues. In addition, he said, although engineers can teach micro-ethics easily, they lack the competence to teach macro-ethics. Indeed, he remarked, engineering students may regard the entire subject of ethics skeptically, given what Daniel described as a world around them that can at times appear alarmingly unethical.

To overcome that skepticism, Daniel suggested engineers should organize to promote a Good Samaritan law to protect engineers who come to the aid of people in a time of crisis. In support of this idea, he described his experience as chair of the American Society of Civil Engineers Hurricane Katrina external review panel. In his opinion, things had gone wrong when the levees were built in New Orleans because the focus of those promoting and implementing the project was on profit rather than on social justice. Too little money was spent on flood protection to ensure safety and social justice, he said, and there was a general unwillingness among those in a position to call the country's attention to the needs to spend political capital to change that. As a result, the engineering of the levees was poor. However, it is too late to correct that situation. All that can be done in the near term is to focus on evacuation plans, he said.

Dr. Daniel concluded with a reminder that the audience should consider which aspects of ethics and social justice should be taught in college classes and which should be taught in continuing education. Given the time constraints in undergraduate engineering education, this is an important issue.

5.5 GENERAL DISCUSSION

Eric Pappas, school of engineering at James Madison University, noted that the communications and design program at his school is an example of a program that crosses disciplinary boundaries. It includes creative thinking, aesthetics, and ethics, among other topics.

The audience then turned quickly to an intense discussion of the meaning of "social justice." Some argued that social justice is a political concept, others that it has religious connotations; some considered relative risk in relationship to social justice, arguing that the most vulnerable among us are often exposed to disproportionately greater risk. Still others said they considered ecological sustainability a social justice issue. Many agreed that technical knowledge is essential to understanding the social and ethical parameters of choices, whether the choices relate to transportation or fisheries or any other social activity. Engineers, they argued, should be involved in educating the public about engineering and social justice issues, but they questioned whether their education enabled them to assume that role (although one said that the Tufts program was an exemplar of how this could be done). Participants disagreed about how well engineering societies are educating the public, and even about whether engineering societies should be speaking with one voice about ethical matters.

The group also discussed whether engineering educators realize that most students do not pursue academic careers and whether they should be preparing students to deal with ethical issues that arise in non academic environments. Dr. Passino had the last word on this subject. Since only about 15 percent of engineering undergraduates at Ohio State go on to graduate school, he said, he designed a less theoretical course that focuses on issues like safety and risk in manufacturing, a situation to which many students will actually be exposed. He ended with a plea for a grassroots push for strengthening corporate citizenship programs, which could raise the status of the profession.

6

Small Group Meetings

The workshop was designed to allow time for small group discussions. Four small group meetings were held after Session III on the first day and on the morning of the second day. The first two groups were asked the same question: in regard to ethics, sustainability, and social justice in the engineering profession, how do we get "there"? Group 1 approached the subject from the engineers' perspective, and Group 2 approached it from social perspectives. Group 3 was asked to discuss new approaches in engineering ethics research and education. These groups were asked to use workshop presentations and discussions as a basis for their deliberations. Group 4 discussed how the National Academy of Engineering (NAE) and professional engineering societies might encourage efforts by engineers to promote social justice and sustainable community development. Participants attended one of the four groups. Afterward, rapporteurs for each group described highlights of their discussions in a full plenary session.

The rapporteurs' reports follow; for the names of group leaders, see the agenda and the list of participants in Appendixes A and B. The opinions in the following descriptions were expressed in one or more of the discussions, but they are not consensus opinions nor are they the opinions of the sponsors.

All groups agreed that the words "ethics," "sustainability," and "social justice" are difficult to define, and much time was spent talking about what they mean in an engineering context and how they should be interpreted. Indeed, the issue of defining the terms of the workshop took center stage, and each group attempted to come up with its own definitions. Plenary discussions following the summaries by the rapporteurs revealed sharp differences in opinion about the concept of social justice and its implications, including its implications for engineering. Despite these differences, one group put forth the following "declaration for engineering:"

Engineers and engineering societies have a heritage of concern for ethics and ethical issues. Yet in fulfilling its professional responsibilities, engineering has for too long neglected questions about social justice and sustainable community development. As in other professions, engineers are obligated to serve the public interest. To honor this commitment to public service, engineers should pay greater attention to social justice and sustainable community development. In this way, engineering can take a leadership role in developing a vision of a profession that provides integrated solutions.

6.1 GROUP 1, ENGINEERS' PERSPECTIVES

Karen Smilowitz, Northwestern University, rapporteur of Group 1, first noted that the difficulty of defining terms, even among like-minded people with common concerns, is indicative of their complexity. Group 1 ultimately decided that, for their purposes, getting to social justice requires "achieving equality in human rights" and "equity in human opportunity." The group then moved on to a discussion of how to reach that goal. They identified the following essential participants in that effort: academia (at both the undergraduate and graduate levels), professional organizations, industry, government (local, state, and federal), and nongovernmental organizations (NGOs).

Group 1 discussed how engineers, engineering societies, and employers could identify pathways, solutions, and opportunities to promote just and sustainable applications in engineering and societal systems. If engineers are engaged early in the development of an engineering solution to a problem, they can do more than provide sound engineering advice; they can question whether social justice and sustainability goals are also being met by that particular project. They can also propose sustainable solutions, not just with a 12-month horizon, but with a 10- or 20-year horizon to encompass the long-term social impacts of their work.

> "[Y]ou don't just go into a community and say, 'Here's the solution, see you later,' but make sure that these are sustainable solutions; this is one of the keys to sustainability."
>
> Karen Smilowitz,
> Northwestern University

Engineers should work with social scientists, group members said, to ensure that they are really "listening" to the populations they intend to serve. They have an opportunity and perhaps a responsibility to engage "customers," expand choices, and empower these customers, as was done in the project in Kenya described in the panel on "early career" engineers. This particular project was undertaken with the cultural conditions and needs of the end users taken into account. Engineers should also continually educate the populations affected by the engineering solutions they provide. Performance and economic goals should not be the only ones taken into account in determining the success of a project—environmental, ecosystem, and social equity goals should all be considered. Project designs should also include a commitment to outcomes assessment and data gathering to determine the effects on the population.

The group concluded that academia is not currently structured to reward interdisciplinary work—which is crucial to developing curricula and programs on ethics and social justice. The group suggested that the tenure process and academic funding structures be altered to take into account the importance of interdisciplinary efforts.

Another way to publicize and legitimize these efforts to traditional practitioners of engineering is to encourage leading technical journals to feature humanitarian work. Disseminating and publicizing the social justice and sustainability aspects of engineering projects may also tap into the enthusiasm of many students for making a difference and show that there is a place for them to call "home" once they graduate. If the market potential of "sustainable development" can be demonstrated, this would tap into students' entrepreneurial spirit. Universities must take the lead in broadening the student knowledge base, giving them multidisciplinary exposures, adjusting teaching techniques, and focusing more on the humanitarian aspects of engineering solutions to problems. Schools can also promote opportunities for students to work in the real world with NGOs, the United Nations, and similar organizations through summer- and semester-long cooperative learning programs. The group also noted that engineering societies should lead the way by adopting a coherent message and providing continuing education opportunities.

6.2 GROUP 2, SOCIAL PERSPECTIVES

Michael Loui, University of Illinois Urbana-Champaign, presented the summary for Group 2, which discussed how engineering projects (e.g., interstate highways that devastate minority communities in urban areas, flood levees in New Orleans) can have unintended social consequences. These negative consequences might be a function of engineers operating far from decision makers and with "blinders" on in large organizations such as the Army Corps of Engineers, the National Aeronautics and Space Administration, and others in the private sector, that have been shown to ignore the social and, often, the environmental consequences of their work. When faced with these problems, individual engineers need and deserve assistance; professional societies should assume responsibility for questioning the structure and functioning of these large organizations.

Several members of Group 2 argued that engineers should feel compelled to think about long-term consequences and sustainability. Even if this is not an assigned part of their jobs, they should feel an individual professional responsibility that is an integral aspect of who they are as engineers. It is true that engineers often work under constraints set by others, but they should also think about their tasks in a sociotechnical context and focus on the ends as well as the means to achieving their goal.

> "[O]ften engineering work is divorced from the social context."
>
> Michael Loui,
> University of Illinois,
> Urbana-Champaign

Along the same lines, the group urged that, although engineers cannot take full responsibility for setting social goals, they and their societies should strive to increase awareness of these issues among engineers and to develop a process of engagement with clients, employers, and the public to discuss the implementation of their work in a responsible way.

The ethos of the engineering profession should encourage engineers to contribute to social justice and sustainability. This ethos should be so intertwined with the profession that it becomes common for employers to support efforts by their employees to undertake projects in pursuit of these goals. Codes for the profession should recognize this obligation.

Engineers and engineering societies should engage in an intense dialogue with scholars in the humanities and social sciences on how engineering can serve the cause of social justice. In turn, members of those disciplines should engage with engineers to determine how they can help them work toward these goals. Engineers should also engage in a public conversation on social justice issues and ensure that representatives of typically disadvantaged groups are involved.

6.3 GROUP 3, ETHICS EDUCATION

Bruce Seely, Michigan Technological University, spoke for the third breakout group, which was asked to identify successful programs that address social justice and sustainability issues. The group began by praising many of the programs with which the workshop participants were affiliated. They then identified other programs around the country that focus on social justice, sustainability, and related topics. The more important question, the group suggested, is whether these programs have moved beyond individual initiatives to gain institutional support and whether they can be reasonably exported and duplicated.

The question of "who should teach what" must be addressed when an institution is considering whether, and how to incorporate social justice and humanitarian concerns into the curriculum. Many engineering instructors are not well prepared to teach ethics in a substantive, professional way, which raises an important question—should there be new required courses, with the involvement of humanities and social science faculty, or, perhaps, ethics material be integrated into courses throughout the curriculum?

To define "ethics" in this context, the group decided they needed more information. For instance, what is the appropriate level for classroom discourse? Discussions could include meta-ethics (looking for conceptual clarification); normative ethics (which, like practical, professional, and applied ethics, focuses on ethical implications of particular problems); paramedic ethics (encouraging professionals to use rules of thumb and standards of practice to address ethical issues); and even pop ethics (online blogs by professionals and members of the public who express their opinions on a wide variety of ethical issues in science and engineering).

Understanding the distinctions among these approaches is crucial to teaching ethics, they said. Teaching in context is an important aspect of all successful engineering instruction, but teaching ethical engineering "in context" will require adjusting materials and discussions to engineering students (e.g., students working in a lab) who may not recognize, or may even flatly deny, that there are ethical considerations in their work.

The group then turned to a discussion of the purpose of introducing ethics in engineering education in the first place. Since many people understand the term "ethical," as in "ethical engineer," to mean a person who follows general moral precepts, some engineers may be uncomfortable with ethical precepts as a professional goal. Several group members noted that introducing and improving ethical reasoning, particularly in professional contexts, might be more easily accepted.

The concept of social justice, which addresses issues in social and economic contexts, often includes "ethics" but is more political. Ultimately, some argued, the common characteristic of sustainability and social justice is "values," which, perhaps, should be introduced to students as a design parameter. Engineers should listen, rather than impose; end the expert-user divide; and take into account local and indigenous knowledge in making engineering decisions.

Successful programs in ethics bring critical thinking into the process in a variety of ways, such as asking new and different kinds of questions. Therefore, it is important to introduce a variety of perspectives on how to address problems, from feminism to race and ethnic studies to postcolonial studies, and so on.

Group 3 also took up some broad issues related to bringing ethics into engineering education, such as reexamining structures in the curriculum that lead students to feel "they must cheat to survive." The group considered whether the "first" engineering degree should be a six-year master's degree, that perhaps the four-year curriculum is too crowded and too short for the complete development of engineering students. They drew a distinction between training and education. Engineering programs should ask themselves whether they are producing "technicians" or "professional engineers."

Finally, the group discussed how new media could be used in sophisticated ways to transmit the codified information in engineering curricula. This would leave more time for the discussion of issues of societal and ethical importance. Given that there are few models to follow, the best advice now, the group suggested, is to "do whatever you can make work."

In considering the role of the Center for Engineering, Ethics, and Society (CEES), the group noted that it would be helpful to know how deeply engaged the NAE membership is in these issues. How many would be willing to participate in disseminating and supporting the ideas generated by CEES? Perhaps CEES could focus on developing a portfolio of cases to assist engineering faculty, who often have very limited preparation for teaching ethics. In addition, CEES could lead an effort at the graduate level to ensure that potential new faculty members are familiar with ethical materials and issues before they become instructors. CEES could also provide assistance in engaging social science and humanities faculty.

Many in the group felt that there might be an emerging professional identity for instructors who teach engineering ethics, which raises a number of questions CEES might investigate. Is this a new branch of engineering? If so, what should it be called and what would its curriculum include? How would a major in this area affect a career? In an educational environment that values publications, research, and grants, how could the risks of someone getting "stuck" with no clear path to tenure be mitigated?

The answers to these questions will not be the same for all schools. Even at large research universities, where NSF efforts in grant making have made some inroads in introducing ethics, there is still little tolerance for a career path based on ethics. Smaller schools, however, might be more open to introducing new perspectives to their faculty. The approach would vary, depending on the audience, as will the response.

6.4 GROUP 4, PROFESSIONAL SOCIETIES

Joe Herkert, Arizona State University, Polytechnic Campus, was the rapporteur for Group 4, which focused on what NAE and other professional societies can do. The group suggested that NAE attempt to define the meaning of "social justice" in the context of engineering and then provide a voice for the profession with an appropriate message. The American Association for Engineering Societies (AAES) and a number of other societies have established forums on sustainability, and it would be useful to expand on those efforts, with NAE presenting the idea to their boards (based on the assumption that leadership must be engaged before members are likely to become involved). NAE can, perhaps, use the annual Convocation of Academic Engineering and Technical Societies to promote this idea.

To encourage changes in professional practices, societies must maintain a highly visible, continuing dialogue on these issues. Members of Group 4 suggested that professional societies carefully develop overt and covert messages on engineering, social justice, and sustainable community development and related issues, such as diversity in engineering. They could sponsor workshops in the corporate world, thus providing venues for sharing best practices and discussing issues of engineering and business ethics, and they could host meetings of business executives to discuss the need for more diversity in engineering and engineering leadership. In addition, professional societies should recognize exemplars with prominent conference sessions, awards, and prizes.

The group discussed how CEES could work with other engineering organizations, such as AAES and the World Federation of Engineering Organizations, to promote international cooperation on sustainability and social justice, as well as a focus on ethical leadership and organizational ethics that go beyond legal requirements.

NAE and other professional organizations could promote career opportunities provided by Engineers Without Borders and Engineers for a Sustainable World. Engineering students, including women and underrepresented minorities, are interested in the kinds of projects undertaken by these organizations, as well as in combining entrepreneurship with social justice and humanitarian goals.

Many in the group argued that professional codes of ethics should be reexamined and modified to encourage sustainability and social justice as professional goals. Codes of ethics might also be improved by explicitly recognizing multiple-value perspectives and expanding the self-policing efforts of engineering professions. Finally, the group noted that ABET could elevate the importance of ethics and sustainability, citing the American Socieety of Civil Engineers "Red Book" as an example. They also argued that more emphasis should be put on reviewing program performance in ethics areas as rigorously as in technical areas and suggested that ABET might revise its general criterion 3f: "Understand professional and ethical issues" to include the phrase "including sustainability and social justice." Sustainability and social justice could become standard parts of continuing education to raise their visibility in the community of professional engineers.

7

Roundtable Discussion: The Intersection of Humanitarian Action, Social Justice, and Sustainable Community Development

In doing their work, engineers may face circumstances where issues of social injustice or environmental vulnerabilities arise, and/or humanitarian actions have to be taken. The subject of the roundtable discussion was how engineering and engineering ethics research, practice, and education could be improved so engineers are better prepared to deal with these circumstances. Presenters and discussants were asked to consider the following questions: "What research and practical efforts are needed? What sources of support for these efforts exist and can be promoted?"

Sheila Jasanoff, Science and Technology Studies, Kennedy School, Harvard University, moderated the session. Presenters were Carlos Bertha, Department of Philosophy, U.S. Air Force Academy; Regina Clewlow, founding director, Engineers for a Sustainable World (ESW), and energy fellow and Ph.D. candidate, Massachusetts Institute of Technology (MIT); and Juan Lucena, Center for Engineering, Ethics, and Society Advisory Group (CEES-AG) member, Liberal Arts and International Studies Division, Colorado School of Mines. The discussants were Garrick Louis, Systems Engineering, University of Virginia; Bill Wallace, Wallace Futures Group and Engineers Without Borders (EWB) International; and Dennis Warner, senior technical advisor, Catholic Relief Services.

7.1 THE TEACHING OF ENGINEERING ETHICS

In the first presentation, Carlos Bertha, Department of Philosophy, U.S. Air Force Academy, addressed the question of who should teach ethics and engineering ethics. He said he could draw on his philosophical training and his current career, as well as his training and professional experience as a mechanical engineer in both design and construction in civilian and military life here in the United States and, most recently, in Afghanistan. He also noted that in talking about community sustainable development and social justice he could draw on his experience as a member of the South Carolina National Guard and the U.S. Northern Command whose Joint Support Group responded to Hurricane Katrina.

Dr. Bertha focused on the teaching of engineering ethics education. He noted that practicing engineers often criticize the philosophical approach to teaching ethics for being too theoretical, while philosophers often criticize applied courses as teaching compliance rather than ethics. He suggested that a casuistical approach, a focus on ethical issues in particular cases, would help improve engineering practice. He argued that philosophers might contribute to engineering courses by providing ethical analyses of cases being developed or used in engineering departments. Philosophers could teach these course components or help engineering faculty teach them.

> "[T]hey [students] need to wrestle with the fact that there are situations for which there are no formulas and that the best you can do is put together a cogent argument."
>
> Carlos Bertha,
> U.S. Air Force Academy

A casuistical approach would require that engineering students recognize that no mathematical formula can provide "the one" solution. However, the correct approach is not "all subjective" either. It entails arguing through the ethical content of a specific case and identifying the morally relevant aspects of the case. A provisional resolution might be developed, but it might have to be changed later.

In his work in Afghanistan, Dr. Bertha said, it was important to guard against paternalism, that is, assuming that the intervening party (in this case, the engineer) knows how to solve the problem (and that the home party wants to be passive). It was also important to guard against moral relativism, for instance, a situation in which the intervening party may condone bribery because it is common practice. Exchanges that educate engineers about cultural differences would help them approach sustainable community development in foreign countries, such as Afghanistan, more effectively, he said. He suggested forming multinational engineering and construction partnerships that include small business partners in underdeveloped countries.

7.2 CURRICULUM-BASED SUSTAINABLE DEVELOPMENT PROJECTS

Regina Clewlow is currently an energy fellow and Ph.D. candidate in engineering management at MIT. She came to the program after spending six years as founding director of ESW. Her classes now include problems of social justice and engineering, management, and society, called "complex engineering systems." The nomenclature may reflect an attempt by engineers to recognize the importance of including social behavior in engineering studies; it also reflects a recognition of connections between engineering, society, public policy, and management.

ESW is a group that engages faculty and students in curriculum-based sustainable development projects around the world and in the United States. For example, the goal of a project might be to improve access to clean water and sanitation, or, in the United States, it might focus on sustainability challenges and promoting alternative energy. Ms. Clewlow challenged her listeners to think about how engineering ethics research and education can improve engineering projects in the contexts of humanitarian crises and social and environmental issues.

ESW, she said, does this in a number of ways. It encourages its campus chapters to link projects to students' grades, which would ensure that they would be attentive to discussions of humanitarian and social justice issues. Plugging projects into the curriculum, she said, also ensures that students will have some oversight, which is important for their protection, as well as for more effective project outcomes. Because projects often entail skills in numerous disciplines, they often require a lead instructor and co-instructor. The most effective projects are launched by universities with strong interdisciplinary programs.

Ms. Clewlow also explained the rationale for involving local partners and undertaking projects that last for a substantial period of time so that local priorities can be well understood and results can be sustained by the communities served by projects. In closing, she reiterated the need for engineering ethics research on effective ways for engineers to engage in sustainable community development and the need for practical resources for the engineering community that decides to engage in such projects. Specific resources depend on project specifics, but all require adequate supervision and multidisciplinary and well-experienced expertise.

7.3 INTEGRATING CRITERIA FOR CONSIDERING HUMANITARIAN ASSISTANCE IN ENGINEERING CURRICULA AND PROFESSIONAL CODES OF ETHICS

The third presentation, by Juan Lucena, CEES-AG member, Liberal Arts and International Studies Division, Colorado School of Mines, was on relationships between engineering and the historical, ideological, and institutional dimensions of humanitarianism and sustainable community development. Dr. Lucena proposed changes in curricula and codes of ethics to prepare engineers to work in circumstances that involve humanitarian and social issues. He contended that engineers must understand how the social history and organizational contexts in which they operate affect the choices they can make and the outcomes likely to come from those choices, as well as how they affect the problem-solving approaches with which they are familiar. Students must recognize that someone with a desire to help often proceeds without questioning whether help is wanted or thinking about how such well-intentioned "help" might provoke a crisis. Based on historical frameworks for humanitarianism, and even sustainable community development, a problematic status quo may sometimes be left as is. For example, if the situation for the hungry may become measurably better by using more environmentally benign agricultural methods, but yet unjust power structures would also be strengthened, it might be less destructive to leave the situation as it is.

Professor Lucena and his colleagues developed the following guidelines for deciding whether to provide humanitarian assistance:

- Will the engineering work promote the good of all people?
- How might this project protect and promote human rights?
- Is the engineering solution likely to solve a humanitarian crisis?
- Will the engineering intervention address fundamental human needs?
- Will it provide benefits for those who are presently underserved?
- Is the work more compatible with not-for-profit than for-profit enterprises?
- What is the likelihood that the product, process, or system will be sustainable?

- Have the cultural exigencies of all stakeholders been taken into account?

> "Quite often the students and practicing engineers do not realize that actually it does make a difference when you come with an imprint of the [U.S.] Agency for International Development or the World Bank, and that makes a difference and creates a power imbalance wherever you are operating. . . . "
>
> Juan Lucena,
> Colorado School of Mines

Successful humanitarian and sustainable community development, Dr. Lucena maintained, requires attention to the social dimensions that influence the successful adoption of a technology; to community capabilities rather than deficiencies; to interrelationships and interdependencies in communities; and to the need for community ownership and buy-in.

These considerations could be incorporated into traditional engineering ethics education and professional codes, and the designers of engineering design projects and service learning projects could assess those projects according to these considerations. For example, the National Society of Professional Engineers might develop a separate code component for engineers working in these contexts. ABET could develop a specific criterion for humanitarian engineering programs, whereby students would be expected to demonstrate an ability to recognize community capabilities and worldviews different from their own. Dr. Lucena suggested that employers of engineers in these contexts could also include that criterion in hiring and decision making. Finally, funding agencies might use such criteria in evaluating the projects they support.

7.4 PANEL DISCUSSION

Garrick Louis, Systems Engineering Department, University of Virginia, opened the discussion with some basic questions about defining "development" and "community." For example, should we assume that our hosts use the same definitions we use? He then raised the issue of the distribution of risks and benefits and posed two further questions, about "process" and "product." Process, he said, is about motivation and the need for the engineering intervention; about how the community will be engaged; and about exit strategies. "Product" is about sustainable results and capacity building and expected research and education products. Process and product need careful thought before people engage in a community development activity, so that whatever is done is acceptable and results can be sustained. Product in particular may have a focus on capacity building, both in communities to be helped and in the students who undertake the projects. For faculty, product may also include the potential for publications and grants.

Dr. Louis noted that there are a variety of imperatives for outsiders entering into communities. They may be responding to moral imperatives or to their interest in promoting new social or technical knowledge. He argued that successful interventions require a focus on capacity development, leaving communities with the capabilities to handle the technical, social, and financial aspects associated with project assistance, and even with the ability to teach other communities what they have learned.

Bill Wallace, of Wallace Futures Group and EWB International, stressed the urgency of addressing the problems facing the world today—particularly in environmental problems. Sustainability is a matter of necessity, he said, as two recent books had made clear, *High Noon: 20 Global Problems, 20 Years to Solve Them* by J. F. Rischard (Basic Books, 2003) and *Capitalism as if the World Matters* by J. Porritt (Earthscan, 2006). Dr. Wallace said he expected to take information criteria and checklist items (particularly the need for collaboration and careful listening) that had been presented during the workshop back to EWB to use in planning future projects.

Before turning to the final discussant, session moderator Sheila Jasanoff, Science and Technology Studies, Kennedy School, Harvard University, noted that the effectiveness of aid in an emergency may depend greatly on understanding the psychology and needs of the community. As an example, she cited the response of a fundamentalist mission to India that brought popular sweets to people in the aftermath of a disaster. The recipients appreciated the gift, while ignoring the team from Doctors Without Borders that was trying to meet their medical needs but didn't know how to engage their trust.

The final discussant, Dennis Warner, Senior Technical Advisor, Catholic Relief Services, reviewed the strengths and weaknesses of engineering approaches to humanitarian crises, focusing on why current approaches, driven by market and commercial forces, pose challenges to engineering professionalism. Mr. Warner proposed that a branch of engineering be established, called "humanitarian engineering," that would focus on engineering interventions to address injustices and improve necessary services for people in need. He highlighted six areas in which practices and guidelines toward this end might be improved: university education, volunteer service, codes of ethics, legislation for national service, private sector support, and media promotion.

7.5 GENERAL DISCUSSION AND WRAP-UP

In the general discussion at the end of the meeting, participants invited each other to get acquainted with resources that they believed would be helpful in enhancing engineering for social justice and sustainable community development or in providing venues for publication of their efforts and the results of relevant research. The philosophers mentioned were Bernard Williams, Alisdair MacIntyre, and Annette Baier. Among the resources were the Online Ethics Center at NAE at *onlineethics.org* and the IEEE journal *Technology and Society*. Participants emphasized new themes for teaching engineering ethics raised in workshop presentations and discussion: taking a systems-oriented approach that includes understanding its social dimensions, increasing the use of critical thinking and the social sciences (especially science and technology studies); and training in the development of empathy and people skills and attentiveness to local knowledge. People reiterated the need, strongly felt, to become partners with the communities in which service would be offered and to examine cultural priorities including systems of patronage and religious affiliation that could affect understanding and reception of offers of assistance, and also encourage positive social action. In closing, the session moderator pointed out that these points raise further questions, leaving open-ended considerations for continuing the conversation.

Appendixes

Appendix A

Workshop Agenda

ENGINEERING, SOCIAL JUSTICE, AND SUSTAINABLE COMMUNITY DEVELOPMENT

NAS BUILDING
2100 C STREET NW,
WASHINGTON, D.C.

OCTOBER 2, 2008

8:00 a.m. Continental Breakfast

8:30 *Opening Session:*

Moderator:
John Ahearne, Chair, NAE Center for Engineering, Ethics, and Society (CEES) Advisory Group

8:30 Welcome—Charles M. Vest, NAE President

8:45 Laying the Groundwork—William A. Wulf, Past NAE President, Distinguished Professor, University of Virginia

9:00 CEES Meeting Overview and Logistics
John Ahearne, Chair, CEES Advisory Group
Rachelle Hollander, Director, CEES

9:15 ***Session I: Engineering and Special Vulnerabilities***

Engineers and engineering organizations operate in circumstances of crisis, ranging from conflict to disaster. They operate where human rights problems are highly visible, and where issues of sustainable community development arise. This session reports perceptions about the technical and social constraints and opportunities they face, and whether and how aims for humanitarian action, social justice, and sustainable community development can be met.

Moderator:
Henry J. Hatch, NAE Member, Former Commander, U.S. Army Corps of Engineers

Presenters:
Abul Barkat, Economics, University of Dhaka, Bangladesh
Abul Hussam, Chemistry, George Mason University, and winner of the NAE Grainger Challenge Prize for Sustainability
Christopher Seremet, Technical Advisor, Catholic Relief Services
Anu Ramaswami, Civil Engineering, University of Colorado, Denver

Discussants:
George Bugliarello, NAE Member, University Professor, Polytechnic University
Deborah Goodings, Engineering and Public Policy, University of Maryland, College Park

11:00 ***Break***

11:15 ***Session II: Engineering, Ethics, and Society***

This session explores the interface of engineering, ethics, and practice. Do humanities and social sciences disciplines bear on problems for engineers and engineering professions in such circumstances? Has the field of engineering ethics drawn adequately from this scholarship or the real exigencies of engineering practice? This session will examine technical, political, historical, environmental, economic, and cultural constraints that shape outcomes.

Moderator:
David Crocker, Institute for Philosophy and Public Policy, University of Maryland, College Park

Presenters:
Ron Kline, Science and Technology Studies, Cornell University
Carl Mitcham, Philosophy, Colorado School of Mines
Wesley Shrum, Sociology, Louisiana State University

Discussants:
Priscilla Nelson, Provost and Vice President for Academic Affairs, New Jersey Institute of Technology
Donna Riley, Science and Environmental Policy, Smith College

12:45 p.m. ***Lunch***

1:45 ***Early Career Engineers Panel***

What led you to your career choices? How do you see them in relationship to the goals of this meeting, to enhance engineering research and practice and improve engineering education through attention to issues of engineering, social justice, and sustainable community development?

Moderator:
Joseph Le Doux, Biomedical Engineering, Georgia Institute of Technology

Panelists:
Rebekah Green, Institute for Global and Community Resilience, Western Washington University
Daniele Lantagne, Centers for Disease Control and Prevention

2:30 ***Session III: Implications for Engineering Education***

Engineering educators and professional societies can promote attention to engineering in circumstances of social conflict and environmental challenge, and prepare students and members to address issues responsibly. What kinds of challenges do such activities bring to engineering education? How are educational and professional programs responding? What structural, programmatic, and/or curricular changes, if any, are underway? What alliances are needed?

Moderator:
Woodie Flowers, NAE Member, Mechanical Engineering, Massachusetts Institute of Technology

Presenters:
Linda Abriola, NAE Member, Dean, School of Engineering, Tufts University
Caroline Baillie, Materials Engineering & Engineering Education, Queens University, Canada
Kevin Passino, Electrical and Computer Engineering, Ohio State University

Discussants:
Richard Anderson, SOMAT Engineering, Inc.
David Daniel, NAE Member, President, University of Texas, Dallas

3:45 ***Break***

4:00 ***Small Group Sessions***

Participants will deliberate on interventions and actions to enable engineers and organizations to develop and assess strategies for change as well as make proposals for implementation.

5:30 ***Adjourn for the day***

OCTOBER 3, 2008

8:00 a.m. Continental Breakfast

8:30 *Small Group Sessions II: Prepare for Plenary*

9:30 *Plenary*
Presentation and discussion of the small group reports will be followed by general discussion and suggestions for next steps.

11:00 *Roundtable: Addressing the Intersection of Humanitarian Action, Social Justice, Sustainable Community Development*
This roundtable will consider how engineering and engineering ethics research, practice, and education might better address the complex choices and cultural conflicts facing engineering in these circumstances. What research and practical efforts are needed? What sources of support for these efforts exist and can be promoted?

Moderator:
Sheila Jasanoff, Kennedy School, Harvard University

Presenters:
Carlos Bertha, Philosophy, U.S. Air Force Academy
Regina Clewlow, Founding Director, Engineers for a Sustainable World
Juan Lucena, Liberal Arts and International Studies, Colorado School of Mines

Discussants:
Garrick Louis, Systems Engineering, University of Virginia
Bill Wallace, Wallace Futures Group and Engineers Without Borders, International
Dennis Warner, Senior Technical Advisor, Catholic Relief Services

1:00 p.m. *Lunch, Informal Review of Results, and Next Steps*

2:15 *Workshop adjourns*

Small Group Sessions and Topics

Group 1: Engineering for Social Justice and Sustainable Community Development—how to get there: engineers' perspectives. This group will assess the first session, identifying the main assumptions and findings, as well as what was overlooked.

Group Leader:
Jimmy Smith, National Institute for Engineering Ethics, Texas Tech University

Rapporteur:
Karen Smilowitz, Industrial Engineering and Management Sciences, Northwestern University

Group 2: Engineering for Social Justice and Sustainable Community Development—how to get there: social perspectives. This group will assess the second session.

Group Leader:
Ed Harris, Philosophy, Texas A&M University

Rapporteur:
Michael Loui, ECE, University of Illinois, Urbana-Champaign

Group 3: New approaches in engineering ethics research and education. This group assesses session three.

Group Leader:
Keith Miller, EECS, University of Illinois, Springfield

Rapporteur:
Bruce Seely, Social Sciences, Michigan Technological University

Group 4: Roles for NAE and engineering societies.

Group Leader:
Alice Agogino, NAE Member, Mechanical Engineering, University of California, Berkeley

Rapporteur:
Joseph Herkert, Technology and Ethics, Arizona State University, Polytechnic Campus

Appendix B

Workshop Attendees

Linda Abriola* (NAE)
Dean, School of Engineering
Tufts University
105 Anderson Hall
200 College Ave.
Medford, MA 02155
617.627.3237
engineeringdeansoffice@tufts.edu

Alice Agogino* (NAE)
Roscoe and Elizabeth Hughes Professor of Mechanical Engineering
University of California at Berkeley
5136 Etcheverry Hall
Mailstop 1740
Berkeley, CA 94720
510.642.6450
agogino@berkeley.edu

John Ahearne* (NAE)
Director, Ethics Programs
The Sigma Xi Center
P.O. Box 13975
99 Alexander Drive
Research Triangle Park, NC 27709
919.547.5213
ahearne@sigmaxi.org

Adjo Amekudzi
Associate Professor, School of Civil and Environmental Engineering
Georgia Institute of Technology
790 Atlantic Drive
Atlanta, GA 30332
404.894.0404
adjo.amekudzi@ce.gatech.edu

*Program participant.

Richard O. Anderson*
Principal Engineer
SOMAT Engineering, Inc.
26445 Northline
Taylor, MI 48180
734.946.4966
Roape1@aol.com

Nickie Athanason
1622 N. Highland Street
Arlington, VA 22201
ndymersky@hotmail.com

Caroline Baillie*
DuPont Chair in Engineering Education Research and Development
Queen's University
Beamish-Munro Hall, Room 301
Kingston, ON K7L 3N6
613.533.6249
cbaillie@post.queensu.ca

Sarah Banas
Program Associate
American Association for the Advancement of Science
1200 New York Avenue, NW
Washington, DC 20005
202.326.6492
sbanas@aaas.org

Abul Barkat*
Professor, Faculty of Social Sciences
University of Dhaka, Bangladesh
Ramna, Dhaka—1000
Bangladesh
880.2.9661920.73 Ext. 6451
hdrc@bangla.net

Rosalyn W. Berne
Associate Professor, School of Engineering
University of Virginia
351 McCormick Road
Thornton Hall A236
Charlottesville, VA 22904
434.924.6098
rwb@virginia.edu

Kristen Sanford Bernhardt
Assistant Professor, Department of Civil and Environmental Engineering
Lafayette College
310 AEC
Easton, PA 18042
610.330.5584
sanfordk@lafayette.edu

Carlos Bertha*
Associate Professor of Philosophy
U.S. Air Force Academy
HQ USAFA/DFPY
2354 Fairchild Drive, Suite 1A45
U.S. Air Force Academy, CO 80840
719.333.8655
carl.bertha@usafa.edu

Jason Borenstein
Director of Graduate Research Ethics Programs
School of Public Policy
Georgia Institute of Technology
685 Cherry Street
Atlanta, GA 30332
404.385.2801
jason.borenstein@pubpolicy.gatech.edu

Danielle Boykin
Associate Editor
National Society of Professional Engineers
1420 King Street
Alexandria, VA 22314
703.684.2879
dboykin@nspe.org

William Brinkman
Assistant Professor, Department of Computer Science
Miami University
Benton Hall, Room 205
Oxford, OH 45056
513.907.7165
brinkmwj@muohio.edu

George Bugliarello* (NAE)
President Emeritus and University Professor
Polytechnic Institute of NYU
6 MetroTech Center
Brooklyn, NY 11201
718.260.3330
gbugliar@poly.edu

Vincent Caristo
Undergraduate Student, School of Engineering
University of Virginia
138 Goodman Street
Charlottesville, VA 22902
609.731.6129
vrc9u@virginia.edu

Dennis Chamot
Associate Executive Director, Division on Engineering and Physical Sciences
The National Academies
500 Fifth Street, N.W.
Washington, DC 20001
202.334.1355
dchamot@nas.edu

Regina Clewlow*
Energy Fellow
MIT Energy Initiative
77 Massachusetts Avenue
E40-455
Cambridge, MA 02139
617.997.1922
rclewlow@mit.edu

Laura Cooney
Assistant Program Manager
Air Force Office of Scientific Research
875 N. Randolph Street
Arlington, VA 22203
703.696.9517
laura.cooney@afosr.af.mil

David Crocker*
Senior Research Scholar, Institute for Philosophy and Public Policy
University of Maryland
2101 Van Munching Hall
College Park, MD 20742
301.405.4763
dcrocker@umd.edu

Gary Cruz
Assistant Director of Programs
SHPE/AHETEMS
416 Yates Street
P.O. Box 19019
Arlington, TX 76019
817.272.1116
gary.cruz@shpe.org

John Daly
Coordinator
Development Gateway
14205 Bauer Drive
Rockville, MD 20853
301.460.6364
John.Daly@gmail.com

David Daniel* (NAE)
President
University of Texas at Dallas
800 West Campbell Road
Richardson, TX 75080
972.883.2201
dedaniel@utdallas.edu

Jonathan Essoka
Environmental Engineer/Scientist
U.S. Environmental Protection Agency
1650 Arch Street
3WP10
Philadelphia, PA 19103
215.814.5774
essoka.jonathan@epa.gov

Ana Ferreras
Program Officer
The National Academies
500 Fifth Street, N.W.
Washington, DC 20001
202.334.1697
aferreras@nas.edu

Kevin Finneran
Editor-in-Chief
Issues in Science and Technology
500 Fifth Street, N.W.
Washington, DC 20001
kfinnera@nas.edu
202.965.5648

Woodie Flowers* (NAE)
Pappalardo Professor of Mechanical Engineering
Massachusetts Institute of Technology
77 Massachusetts Avenue
Room 3-43
Cambridge, MA 02139
617.253.6234
flowers@mit.edu

Amy Forrester
Managing Editor
Sustainability: Science, Practice, & Policy
7200 Wisconsin Avenue
Suite 601
Bethesda, MD 20814
301.961.6722
amy.forrester@proquest.com

Deborah Goodings
Professor, Geotechnical Engineering Program
University of Maryland
Department of Civil Engineering
College Park, MD 20742
301.405.1960
goodings@umd.edu

Albert Grant
Chair
Engineers Forum on Sustainability
9036 Willow Valley Drive
Potomac, MD 20854
301.340.8082
aagrant@localnet.com

Rebekah Green*
Research Associate and Grant Writer
Institute for Global and Community Resilience
Western Washington University
516 High Street
Environmental Studies, MS 9085
Bellingham, WA 98225
360.650.4337
rebekah.green@wwu.edu

William Gross
Professor and Dean Emeritus
University of New Mexico
1401 Las Lomas Road N.E.
Albuquerque, NM 87106
505.242.5785
wgross@unm.edu

Robert Hannemann
Director, Gordon Institute
Tufts University
200 Boston Avenue
Suite 2400
Medford, MA 02155
617.627.3110
robert.hannemann@tufts.edu

Muhammad Haris
Professor, Philosophy Department
University of North Carolina, Charlotte
9201 University City Boulevard
Charlotte, NC 28223
704.687.2240
M.Haris@uncc.edu

Ed Harris*
Sue and Harry Bovay Professor of History and Ethics of Professional Engineering
Texas A&M University
302G Bolton Hall
College Station, TX 77843
979.845.5697
e-harris@philosophy.tamu.edu

Hank Hatch* (NAE)
Former Chief of Engineers
U.S. Army
2715 Silkwood Court
Oakton, VA 22124
703.476.8895
hankhatch@aol.com

Joe Herkert*
Lincoln Associate Professor of Ethics and Technology
Arizona State University
Santa Catalina (SANCA) 250D
7001 E. Williams Field Road
Mesa, AZ 85212
480.727.1548
joseph.herkert@asu.edu

Laura Hermann
Director
Potomac Communications
1133 20th Street, N.W.
Suite 400
Washington, DC 20036
202.380.6835
lhermann@pcgpr.com

Robert Hershey
Consultant
Robert L. Hershey, P.E.
1255 New Hampshire Avenue
#1033
Washington, DC 20036
202.659.9529
hershey@cpcug.org

Jonathan Herz
Architect
U.S. General Services Administration
1800 F Street, N.W.
Washington, DC 20405
202.501.3476
jonathan.herz@gsa.gov

Abul Hussam*
Professor, Department of Chemistry and Biochemistry
George Mason University
Fairfax, VA 22030
703.993.1085
ahussam@gmu.edu

Margaret Huyck
Professor
Illinois Institute of Technology
1718 E. 55th Street
Chicago, IL 60615
773.667.3649
huyck@iit.edu

Sheila Jasanoff*
Pforzheimer Professor of Science and Technology Studies
John F. Kennedy School of Government
Harvard University
79 JFK Street, Littauer 354
Cambridge, MA 02138
617.495.7902
sheila_jasanoff@harvard.edu

Jens Kabo
Doctoral candidate
Queen's University, Canada
420 Main Street
Apartment 81
Agawam, MA 01001
413.563.8630
jens.kabo@chee.queensu.ca

William Kelly
Manager of Public Affairs
American Society for Engineering Education
1818 N Street, N.W.
Suite 600
Washington, DC 20036
202.331.3537
w.kelly@asee.org

Samuel Keno
Vocational Training Centers Coordinator
Ethiopia Evangelical Church Development Program
P.box 81064
Addis Ababa, Ethiopia
25.191.135.6062
ksaamii@yahoo.com

Ron Kline*
Bovay Professor in the History and Ethics of Professional Engineering
Cornell University
394 Rhodes Hall
Ithaca, NY 14853
607.255.4307
rrk1@cornell.edu

Margo Kulkarni
Undergraduate Student
Columbia University
1604 W. Alabama
Houston, TX 77006
713.732.6956
mk3046@columbia.edu

Daniele Lantagne*
Safe Water System Staff Engineer
Centers for Disease Control and Prevention
1600 Clifton Road N.E.
MS-A38
Atlanta, GA 30333
404.639.0231
dlantagne@cdc.gov

Joseph Le Doux*
Associate Professor, Department of Biomedical Engineering
Georgia Institute of Technology
313 Ferst Drive, Suite 2127
Atlanta, GA 30332
404.385.0632
joe.ledoux@bme.gatech.edu

Jon Leydens
Associate Professor
Colorado School of Mines
301 Stratton Hall
Golden, CO 80401
303.273.3180
jleydens@mines.edu

Andrew Light
Director, Center for Global Ethics
George Mason University
1736 Columbia Road, N.W.
Apartment 310
Washington, DC 20009
917-570-3276
alight1@gmu.edu

Marshall Lih
Senior Advisor
National Science Foundation
4201 Wilson Boulevard
Room 565.43
Arlington, VA 22230
703.292.4608
mlih@nsf.gov

Michael Loui*
Professor of Electrical and Computer Engineering
University of Illinois
Coordinated Science Laboratory
1308 W Main Street
Urbana, IL 61801
217.333.2595
loui@illinois.edu

Garrick Louis*
Associate Professor, Environmental and Water Resources Engineering
The University of Virginia
Olsson Hall, Office: 102A
151 Engineer's Way
Charlottesville, VA 22904
434.982.2742
gel7f@virginia.edu

Juan Lucena*
Associate Professor, Liberal Arts and International Studies Division
Colorado School of Mines
Stratton Hall #424
Golden, CO 80401
303.273.3564
jlucena@mines.edu

Carl Mack
Executive Director
National Society of Black Engineers
205 Daingerfield Road
Alexandria, VA 22314
703.549.2207
rcox@nsbe.org

Guru Madhavan
Mirzayan Fellow
The National Academies
500 Fifth Street, N.W.
Washington, DC 20001
607.768.4052
gmadhavan@gmail.com

Britney McCoy
Doctoral Candidate
Carnegie Mellon University
5710 Phillips Avenue
Apartment A-1
Pittsburgh, PA 15217
412.521.4546
bmccoy@andrew.cmu.edu

Keith Miller*
Professor of Computer Science
University of Illinois at Springfield
One University Plaza
UIS, CSC, UHB 3100
Springfield, IL 62703
217.206.7327
miller.keith@uis.edu

Carl Mitcham*
Director, Hennebach Program in the Humanities
Colorado School of Mines
Stratton Hall #318
Golden, CO 80401
303.273.3648
cmitcham@mines.edu

Driss Moumane
Senior Advisor
Catholic Relief Services
228 W. Lexington Street
Baltimore, MD 21201
410.625.2220
dmoumane@crs.org

Priscilla P. Nelson*
Provost and Senior Vice President for Academic Affairs
New Jersey Institute of Technology
323 Martin Luther King Jr. Boulevard
Newark, NJ 07102-1982
973.596.3220
pnelson@njit.edu

Dean Nieusma
Assistant Professor
Rensselaer Polytechnic Institute
STS Department, Sage 5508
110 8th Street
Troy, NY 12180
518.276.6381
nieusma@rpi.edu

Douglas Oliver
AAAS Science & Technology Policy Fellow
AAAS (NSF)
1930 N. Van Buren
Arlington, VA 22205
703.292.5112
douglas-oliver@bex.net

Eric Pappas
Associate Professor, School of Engineering
James Madison University
701 Carrier Drive
MSC 4113
Harrisonburg, VA 22807
540.568.1694
PappasEC@jmu.edu

Kevin Passino*
Professor of Electrical and Computer Engineering
The Ohio State University
205 Dreese Laboratory
2015 Neil Avenue
Columbus, OH 43210
614.292.5716
passino@ece.osu.edu

Kurt Paterson
Director, D80 Center
Michigan Technological University
1400 Townsend Drive
Houghton, MI 49931
906.487.3495
paterson@mtu.edu

Greg Pearson
Program Officer
National Academy of Engineering
500 Fifth Street, N.W.
Washington, DC 20001
202.334.2282
gpearson@nae.edu

Iris Portny
Writer/Editor
4300 Old Dominion Drive
#411
Arlington, VA 22207
703.526.4532
iris.portny@cornell.edu

Fionnuala Quinn
12309 Country Ridge Lane
Fairfax, VA 22033
703.758.0851
finnQ@aol.com

Anu Ramaswami*
Professor, Environmental Engineering
University of Colorado, Denver
College of Engineering and Applied Science
Campus Box 113
P.O. Box 173364
Denver, CO 80217
303.556.4734
anu.ramaswami@ucdenver.edu

Donna Riley*
Associate Professor, Picker Engineering Program
Smith College
51 College Lane
Northampton, MA 01063
413.585.7003
driley@email.smith.edu

Jen Schneider
Assistant Professor, LAIS
Colorado School of Mines
1005 14th Street
Stratton Hall 406
Golden, CO 80401
303.273.3628
jjschnei@mines.edu

Brian Schrag
Executive Secretary, Association for Practical and Professional Ethics
Indiana University
618 East Third Street
Bloomington, IN 47405
812.855.6450
appe@indiana.edu

Kenneth Schug
Professor of Chemistry
Illinois Institute of Technology
1466 E. Park Place
Chicago, IL 60637
312.218.3438
kschug@msn.com

Bruce Seely*
Professor of History and Chair, Department of Social Sciences
Michigan Technological University
1400 Townsend Drive
Houghton, MI 49931
906.487.2113
bseely@mtu.edu

Wesley Shrum*
Professor of Sociology
Louisiana State University A&M
126 Stubbs Hall
Baton Rouge, LA 70803
225.388.5102
shrum@lsu.edu

Karen Smilowitz*
Junior William A. Patterson Chair in Transportation
Northwestern University
2145 Sheridan Road
Technological Institute M233
Evanston, IL 60208
847.491.4693
ksmilowitz@northwestern.edu

Jimmy Smith*
Director, National Institute for Engineering
Ethics and Murdough Center for Engineering Professionalism
Texas Tech University
M.S. 41023
Civil Engineering
Lubbock, TX 79409
806.742.3476
Jimmy.Smith@ttu.edu

Bradley Striebig
Associate Professor
James Madison University
3235 Clement Drive
Harrisonburg, VA 22807
540.568.8895
bstriebig@gmail.com

Jessica Tucker
AAAS Science Policy Fellow
Department of Health and Human Services
465 Meadow Road
Apartment 7306
Princeton, NJ 08540
412.654.2065
jessmtuck@gmail.com

S. Claudia Vargas
Director-Founder/Principal Investigator
Complex Systems Optimization Lab
P.O. Box 1087
Northampton, MA 01061
801.362.2862
scvargas@cosola.org

P. Aarne Vesilind
Professor of Civil Engineering (Retired)
Bucknell University
142 Hilltop Place
New London, NH 03257
603.526.6011
aarne.vesilind@gmail.com

Bill Wallace*
Founder & President, Wallace Futures Group
Engineers Without Borders—USA
1400 Overlook Drive
Steamboat Springs, CO 80487
970.879.1122
bill.wallace@wallacefutures.com

Dennis Warner*
Senior Technical Advisor
Catholic Relief Services
228 W. Lexington Street
Baltimore, MD 21201
410.625.2220
dwarner@crs.org

Richard Weibl
Director, Center for Careers in Science
AAAS
1200 New York Avenue, N.W.
Washington, DC 20005
202.326.6674
rweibl@aaas.org

Caroline Whitbeck
Elmer G. Beamer–Hubert H. Schneider Professor in Ethics
Case Western Reserve University
193 Long Bow Lane West
Becket, MA 01223
412.623.5714
cw07@onlineethics.org

Wm. A. Wulf* (NAE)
President Emeritus, NAE
University Professor and AT&T Professor of Engineering and Applied Sciences
University of Virginia
151 Engineer's Way
Charlottesville, VA 22904
434.982.2223
Wulf@virginia.edu

David Zumbrunnen
Warren H. Owen Duke Energy Distinguished Professor
Clemson University
221 Fluor Daniel Building
Clemson, SC 29634
864.656.5625
zdavid@clemson.edu

Secretariat

Rachelle Hollander
Director, Center for Engineering, Ethics, and Society
National Academy of Engineering
500 Fifth Street, N.W.
Washington, DC 20001
202.334.3068
rhollander@nae.edu

Jessica Buono
Research Associate
National Academy of Engineering
500 Fifth Street, N.W.
Washington, DC 20001
202.334.1547
jbuono@nae.edu

Nathan Kahl
Program Associate
National Academy of Engineering
500 Fifth Street, N.W.
Washington, DC 20001
202.334.2226
nkahl@nae.edu

Jacqueline Martin
Senior Program Assistant
National Academy of Engineering
500 Fifth Street, N.W.
Washington, DC 20001
202.334.1628
jmartin@nae.edu